Unlocking Math: A Simple Guide to Complex Problems

Fernando Abednego Halim

© 2024 Fernando Abednego Halim

All rights reserved.

No part of this book may be reproduced, distributed, or transmitted in any form or by any means, including photocopying, recording, or other electronic or mechanical methods, without the prior written permission of the publisher, except in the case of brief quotations embodied in critical reviews and certain other noncommercial uses permitted by copyright law. For permission requests, write to the publisher at the address below:

Fernando Abednego Halim
author.spendable047@passinbox.com

Disclaimer

This book contains AI-generated content. While every effort has been made to ensure the accuracy and completeness of the information, the nature of AI-generated content means that there may be inaccuracies or omissions. Readers are advised to validate the truthfulness and reliability of the content before applying or using it. The authors and publisher disclaim any liability in connection with the use of this information.

ISBN-13: 979-8-3426-4597-3

*Dedicated to all those who continue to ask questions,
embrace challenges, and discover the beauty of mathematics.*

*To my family and friends, whose unwavering support
and encouragement have made this journey possible.*

Contents

Introduction   vii

1 Foundations of Mathematics   1
  1.1 The Language of Mathematics   1
      1.1.1 Basic Symbols and Terminology   1
      1.1.2 Understanding Mathematical Notation   2
  1.2 Arithmetic Essentials   3
      1.2.1 Operations with Whole Numbers   3
      1.2.2 The Importance of Order of Operations (PEMDAS)   3
  1.3 Introduction to Algebra   4
      1.3.1 Variables, Expressions, and Simple Equations .   4
      1.3.2 Solving Basic Algebraic Equations   4

2 Building Blocks of Algebra   6
  2.1 Equations and Inequalities   6
      2.1.1 Linear Equations   6
      2.1.2 Inequalities   7
      2.1.3 Solving Systems of Equations   8
  2.2 Polynomials and Factoring   9
      2.2.1 Understanding Polynomials   9
      2.2.2 Factoring Polynomials   9
  2.3 Quadratic Equations   10
      2.3.1 Solving Quadratics by Factoring   10
      2.3.2 Completing the Square   11
      2.3.3 The Quadratic Formula   11
      2.3.4 Applications of Quadratic Equations   12

3 Exploring Functions — 14
- 3.1 Introduction to Functions — 14
  - 3.1.1 Definition and Notation of Functions — 14
  - 3.1.2 Domain and Range — 15
  - 3.1.3 Types of Functions — 15
- 3.2 Linear and Quadratic Functions — 16
  - 3.2.1 Linear Functions — 16
  - 3.2.2 Quadratic Functions — 17
- 3.3 Advanced Functions — 18
  - 3.3.1 Exponential and Logarithmic Functions — 18
  - 3.3.2 Rational and Radical Functions — 20

4 Geometry Made Simple — 22
- 4.1 Basic Geometric Concepts — 22
  - 4.1.1 Points, Lines, and Angles — 22
  - 4.1.2 Shapes and Their Properties — 23
  - 4.1.3 Properties of Triangles, Quadrilaterals, and Circles — 23
- 4.2 Perimeter, Area, and Volume — 24
  - 4.2.1 Calculating Perimeter and Area of Basic Shapes — 24
  - 4.2.2 Volume of 3D Shapes — 25
- 4.3 Coordinate Geometry — 29
  - 4.3.1 Plotting Points and Lines — 30
  - 4.3.2 Distance and Midpoint Formulas — 30

5 Trigonometry Without Tears — 32
- 5.1 Trigonometric Ratios — 32
  - 5.1.1 Understanding Sine, Cosine, and Tangent — 32
  - 5.1.2 Applications of Trigonometric Ratios in Right Triangles — 33
- 5.2 The Unit Circle — 33
  - 5.2.1 The Unit Circle and Its Significance in Trigonometry — 34
  - 5.2.2 Using the Unit Circle to Find Trigonometric Values — 34
- 5.3 Trigonometric Identities and Equations — 34
  - 5.3.1 Fundamental Identities and How to Use Them — 35

| | | 5.3.2 Solving Basic Trigonometric Equations | 35 |

6 Probability and Statistics — 37
- 6.1 Basics of Probability — 37
  - 6.1.1 Probability Concepts and Rules — 37
  - 6.1.2 Simple and Compound Events — 38
- 6.2 Descriptive Statistics — 39
  - 6.2.1 Measures of Central Tendency — 39
  - 6.2.2 Measures of Spread — 40
- 6.3 Data Interpretation and Analysis — 41
  - 6.3.1 Reading and Interpreting Graphs and Charts — 41
  - 6.3.2 Basics of Statistical Inference — 42

7 Advanced Algebra and Pre-Calculus — 45
- 7.1 Sequences and Series — 45
  - 7.1.1 Arithmetic and Geometric Sequences — 45
  - 7.1.2 Summation Notation and Series — 47
- 7.2 Matrices and Determinants — 48
  - 7.2.1 Introduction to Matrices and Operations — 48
  - 7.2.2 Determinants and Their Applications — 49
  - 7.2.3 Applications of Matrices — 49
- 7.3 Introduction to Limits — 52
  - 7.3.1 Concept of Limits and Their Importance in Calculus — 52
  - 7.3.2 Evaluating Limits Using Algebraic Techniques — 53

8 Introduction to Calculus — 54
- 8.1 The Concept of the Derivative — 54
  - 8.1.1 Understanding Derivatives as Rates of Change — 54
  - 8.1.2 Basic Differentiation Rules — 55
- 8.2 Applications of Derivatives — 56
  - 8.2.1 Derivatives in Motion, Optimization, and Curve Sketching — 56
  - 8.2.2 Related Rates and Real-World Applications — 57
- 8.3 The Integral: Area and Accumulation — 57
  - 8.3.1 Basic Integration Techniques — 57

- 8.3.2 Applications of Integrals in Calculating Areas and Solving Problems of Accumulation 58

9 Problem-Solving Strategies — 61
- 9.1 General Problem-Solving Techniques 61
  - 9.1.1 How to Approach Math Problems Methodically — 61
  - 9.1.2 Common Pitfalls and How to Avoid Them 62
- 9.2 Real-World Math Problems 63
  - 9.2.1 Applying Mathematical Concepts to Real-Life Situations 63
- 9.3 Practice Problems and Solutions 65
  - 9.3.1 Problem 1: Solving Systems of Equations 65
  - 9.3.2 Problem 2: Finding the Derivative 66
  - 9.3.3 Problem 3: Calculating Area Under a Curve 66

10 Tools and Resources — 67
- 10.1 Using Technology in Math 67
  - 10.1.1 Introduction to Calculators, Software, and Online Resources 67
  - 10.1.2 How to Effectively Use Technology to Enhance Learning 68
- 10.2 Additional Learning Resources 69
  - 10.2.1 Books for Further Study 69
  - 10.2.2 Websites and Online Resources 70
  - 10.2.3 Courses and Tutorials 70
  - 10.2.4 Tips for Finding and Utilizing Resources Effectively 71
- 10.3 Preparing for Exams and Beyond 71
  - 10.3.1 Study Strategies for Standardized Tests and College Entrance Exams 71
  - 10.3.2 How to Apply Math Skills in Academic and Professional Settings 72

Conclusion — 74

Introduction

Purpose of This Book

Mathematics is often perceived as a challenging subject, a discipline that requires special talent or innate ability. However, the reality is far different. Math, at its core, is a universal language—a tool that anyone can learn to use effectively with the right guidance. *Unlocking Math: A Simple Guide to Complex Problems* is designed to break down the barriers that many people face when approaching math, transforming complex concepts into simple, manageable steps.

This book is intended for a wide audience: students seeking to improve their math skills, educators looking for effective teaching resources, and lifelong learners who wish to conquer the complexities of mathematics. Whether you are preparing for exams, enhancing your professional skills, or simply seeking to understand the world around you, this book will guide you on your journey.

Overcoming the Fear of Math

The fear of math is a common obstacle that prevents many from reaching their full potential. This book aims to dispel that fear by presenting mathematical concepts in a clear, straightforward manner. Each chapter is crafted to build confidence through understanding, providing you with the tools to approach even the most daunting problems with ease.

Throughout this book, you will find explanations that are not only precise but also relatable, connecting mathematical theory to real-world applications. By bridging the gap between abstract concepts and practical use, this book will show you that math is not something to be feared—it is something to be mastered.

How to Use This Book

Unlocking Math is organized into chapters that progressively build on one another, starting with fundamental concepts and moving towards more advanced topics. Each chapter includes:

- **Clear Explanations:** Every concept is explained in detail, ensuring that you understand the foundational ideas before moving on to more complex material.

- **Practical Examples:** Real-world applications are provided to illustrate how mathematical concepts are used in everyday life and various fields.

- **Visual Aids:** Diagrams and visual representations are included to help you visualize and better understand the material.

- **Practice Problems:** At the end of each chapter, you will find exercises designed to reinforce your understanding and build your problem-solving skills.

Whether you are working through the book from start to finish or focusing on specific chapters that align with your needs, *Unlocking Math* is structured to accommodate your learning style. Use this book as a comprehensive guide, a reference for specific topics, or a tool for review—it's designed to be as flexible as it is informative.

A Note on Citations and References

While this book draws on established mathematical principles and teaching methods, it presents these ideas in a unique way tailored to simplify and clarify complex problems. All original content is carefully crafted to avoid any infringement on copyrighted materials. In cases

INTRODUCTION

where external sources are cited, proper attribution is provided to maintain academic integrity and respect intellectual property rights.

What You Will Learn

By the end of this book, you will have developed a solid understanding of key mathematical concepts, from basic arithmetic to introductory calculus. You will learn to:

- Solve equations and inequalities with confidence.
- Understand and apply functions, from linear to exponential.
- Navigate the world of geometry and trigonometry with ease.
- Analyze data and calculate probabilities.
- Approach calculus with a strong foundation in limits, derivatives, and integrals.

This book is not just about learning math—it's about unlocking the potential within yourself to approach problems with clarity and creativity. The skills you develop here will not only enhance your academic performance but also empower you in your professional and personal life.

Final Thoughts

Mathematics is more than just a subject—it is a way of thinking, a method of understanding the world around us. As you embark on this journey through *Unlocking Math: A Simple Guide to Complex Problems*, remember that each step you take brings you closer to mastery. With dedication and the right guidance, you will find that math is not only approachable but also enjoyable.

Welcome to the world of mathematics. Let's unlock its secrets together.

Chapter 1

Foundations of Mathematics

1.1 The Language of Mathematics

Mathematics is often described as the universal language, a way of expressing ideas that transcends cultural and linguistic barriers. At its core, mathematics is about patterns, relationships, and logical reasoning. To unlock the full potential of math, it's essential to first understand the symbols, terms, and notation that form its foundation.

1.1.1 Basic Symbols and Terminology

Mathematics relies on a set of symbols to convey complex ideas in a concise manner. Here are some of the most common symbols you will encounter:

- $+$ **(Plus)**: Represents addition.
- $-$ **(Minus)**: Represents subtraction.
- $\times$ or $\cdot$ **(Multiply)**: Represents multiplication.
- $\div$ or $/$ **(Divide)**: Represents division.
- $=$ **(Equals)**: Indicates that two expressions are equal.
- $\neq$ **(Not equal)**: Indicates that two expressions are not equal.

- $<$ **(Less than) and** $>$ **(Greater than):** Compare the sizes of two quantities.

- $\leq$ **(Less than or equal to) and** $\geq$ **(Greater than or equal to):** Compare quantities with the possibility of equality.

- $\sqrt{\ }$ **(Square root):** Represents the principal square root of a number.

- π **(Pi):** A constant representing the ratio of a circle's circumference to its diameter, approximately 3.14159.

Familiarizing yourself with these symbols is the first step toward understanding mathematical expressions and equations. As you progress through this book, you'll encounter these symbols frequently, so it's important to recognize them and understand their meanings.

1.1.2 Understanding Mathematical Notation

Mathematical notation is a system of symbols and rules used to represent mathematical concepts and operations. It allows mathematicians to communicate ideas clearly and efficiently. Here are a few examples of how notation is used:

- **Expressions:** A combination of numbers, variables, and operations. For example, $3x + 2$ is an algebraic expression where x is a variable.

- **Equations:** A statement that two expressions are equal. For example, $2x + 3 = 7$ is an equation that can be solved to find the value of x.

- **Inequalities:** A statement that one expression is greater or less than another. For example, $x + 5 > 3$ indicates that x must be greater than -2.

Understanding and using this notation correctly is crucial for solving mathematical problems. As we delve deeper into various mathematical topics, you'll see how these notations are applied in different contexts.

1.2 Arithmetic Essentials

Arithmetic forms the foundation upon which all other branches of mathematics are built. It deals with the basic operations of addition, subtraction, multiplication, and division, and their properties. Mastering these operations is essential for tackling more complex mathematical problems.

1.2.1 Operations with Whole Numbers

Whole numbers are the set of numbers that include zero and all positive integers. The four basic operations—addition, subtraction, multiplication, and division—are fundamental to arithmetic.

- **Addition (+):** Combining two or more quantities. For example, $7 + 5 = 12$.

- **Subtraction (−):** Determining the difference between two quantities. For example, $10 - 3 = 7$.

- **Multiplication (× or ·):** Repeated addition of the same number. For example, $4 \times 3 = 12$.

- **Division (÷ or /):** Splitting a number into equal parts. For example, $20 \div 4 = 5$.

These operations extend to fractions and decimals, where they follow similar rules but require careful attention to detail.

1.2.2 The Importance of Order of Operations (PEMDAS)

When performing arithmetic operations, the order in which you do them matters. The acronym PEMDAS helps us remember the correct order:

- **P: Parentheses** – Perform operations inside parentheses first.
- **E: Exponents** – Solve exponents (powers and roots) next.
- **MD: Multiplication and Division** – Perform from left to right.

- **AS: Addition and Subtraction** – Perform from left to right.

For example, in the expression $3+4\times 2$, multiplication is performed first, so the correct result is $3 + 8 = 11$, not $7 \times 2 = 14$. Following the order of operations ensures that mathematical expressions are evaluated correctly.

1.3 Introduction to Algebra

Algebra is a branch of mathematics that deals with symbols and the rules for manipulating those symbols. These symbols (often letters) represent numbers in equations and expressions, allowing us to solve problems that involve unknown values.

1.3.1 Variables, Expressions, and Simple Equations

A variable is a symbol (usually a letter) that represents an unknown value. In algebra, we use variables to write expressions and equations:

- **Expression:** An algebraic expression combines numbers, variables, and operations. For example, $2x + 5$ is an expression where x is a variable.

- **Equation:** An equation states that two expressions are equal. For example, $2x + 5 = 11$ is an equation that can be solved to find the value of x.

1.3.2 Solving Basic Algebraic Equations

Solving an equation means finding the value of the variable that makes the equation true. Consider the equation $2x + 5 = 11$. To solve for x, follow these steps:

- Subtract 5 from both sides: $2x = 6$.
- Divide both sides by 2: $x = 3$.

1.3. INTRODUCTION TO ALGEBRA

The solution $x = 3$ satisfies the equation, meaning that when x is 3, both sides of the equation are equal.

Algebra provides the tools needed to solve a wide range of mathematical problems, from simple equations to complex functions. As you progress, you'll learn how to apply these tools to increasingly challenging problems.

Recap

In this chapter, we've laid the groundwork for your mathematical journey. We've introduced the basic symbols and terminology of math, explored arithmetic operations, and taken our first steps into the world of algebra. Understanding these foundational concepts is essential as we move forward into more complex topics. Remember, mathematics is a language—once you master its basics, you can unlock its full potential.

Chapter 2

Building Blocks of Algebra

2.1 Equations and Inequalities

Algebra is the language of mathematics, and at its core are equations and inequalities. These are the tools we use to describe relationships between quantities and to solve problems. In this chapter, we'll explore how to solve linear equations and inequalities, and how to apply these skills to more complex situations.

2.1.1 Linear Equations

A linear equation is an equation in which the highest power of the variable is one. Linear equations are among the simplest and most useful types of equations in algebra. The general form of a linear equation is:

$$ax + b = c$$

where a, b, and c are constants, and x is the variable. To solve a linear equation, our goal is to isolate the variable on one side of the equation.

Example: Solve the equation $3x + 4 = 10$

- **Step 1:** Subtract 4 from both sides to begin isolating the variable:
$$3x + 4 - 4 = 10 - 4$$

2.1. EQUATIONS AND INEQUALITIES

$$3x = 6$$

- **Step 2:** Divide both sides by 3 to solve for x:

$$x = \frac{6}{3} = 2$$

The solution is $x = 2$. To verify, substitute $x = 2$ back into the original equation:

$$3(2) + 4 = 10$$

Since both sides of the equation are equal, our solution is correct.

2.1.2 Inequalities

Inequalities are similar to equations but instead of expressing equality, they express a relationship of greater than or less than. The general form of an inequality is:

$$ax + b < c, \quad ax + b > c, \quad ax + b \leq c, \quad \text{or} \quad ax + b \geq c$$

Example: Solve the inequality $2x - 3 > 5$

- **Step 1:** Add 3 to both sides to begin isolating the variable:

$$2x - 3 + 3 > 5 + 3$$

$$2x > 8$$

- **Step 2:** Divide both sides by 2 to solve for x:

$$x > \frac{8}{2} = 4$$

The solution is $x > 4$, meaning that x can be any number greater than 4. Unlike equations, inequalities have a range of possible solutions.

2.1.3 Solving Systems of Equations

Sometimes, we encounter problems involving more than one equation with multiple variables. These are called systems of equations. To solve a system of linear equations, we find the values of the variables that satisfy all the equations simultaneously.
Consider the system:

$$2x + y = 5$$
$$x - y = 1$$

Method 1: Substitution

- **Step 1:** Solve one of the equations for one variable in terms of the other. From the second equation:

$$x = y + 1$$

- **Step 2:** Substitute this expression into the first equation:

$$2(y + 1) + y = 5$$

- **Step 3:** Simplify and solve for y:

$$2y + 2 + y = 5$$

$$3y + 2 = 5$$

$$3y = 3 \quad \Rightarrow \quad y = 1$$

- **Step 4:** Substitute $y = 1$ back into the expression for x:

$$x = 1 + 1 = 2$$

The solution to the system is $x = 2$ and $y = 1$.

2.2. POLYNOMIALS AND FACTORING

Method 2: Elimination

- **Step 1:** Add or subtract the equations to eliminate one of the variables. Add the two equations:
$$(2x + y) + (x - y) = 5 + 1$$
$$3x = 6$$

- **Step 2:** Solve for x:
$$x = 2$$

- **Step 3:** Substitute $x = 2$ back into one of the original equations to solve for y:
$$2(2) + y = 5$$
$$4 + y = 5 \quad \Rightarrow \quad y = 1$$

Again, the solution is $x = 2$ and $y = 1$.

2.2 Polynomials and Factoring

Polynomials are expressions consisting of variables and coefficients, combined using addition, subtraction, and multiplication. The degree of a polynomial is the highest power of the variable in the expression.

2.2.1 Understanding Polynomials

A general polynomial in one variable x is written as:
$$a_n x^n + a_{n-1} x^{n-1} + \cdots + a_1 x + a_0$$
where $a_n, a_{n-1}, \ldots, a_1, a_0$ are constants, and n is a non-negative integer.

Example: $3x^3 - 2x^2 + 5x - 7$ is a polynomial of degree 3

2.2.2 Factoring Polynomials

Factoring is the process of breaking down a polynomial into simpler polynomials (factors) that, when multiplied together, give the original polynomial. Factoring is a key skill in solving polynomial equations.

Example: Factor the polynomial $x^2 - 5x + 6$

- **Step 1:** Find two numbers that multiply to 6 and add to -5. These numbers are -2 and -3.

- **Step 2:** Write the factored form using these numbers:

$$x^2 - 5x + 6 = (x - 2)(x - 3)$$

To verify, expand $(x - 2)(x - 3)$ to confirm that it equals the original polynomial.

2.3 Quadratic Equations

Quadratic equations are polynomials of degree 2, generally written as:

$$ax^2 + bx + c = 0$$

where a, b, and c are constants, and x is the variable.

2.3.1 Solving Quadratics by Factoring

If a quadratic equation can be factored, it can be solved by setting each factor equal to zero and solving for x.

Example: Solve $x^2 - 3x - 4 = 0$

- **Step 1:** Factor the quadratic:

$$x^2 - 3x - 4 = (x - 4)(x + 1)$$

- **Step 2:** Set each factor equal to zero:

$$x - 4 = 0 \quad \Rightarrow \quad x = 4$$

$$x + 1 = 0 \quad \Rightarrow \quad x = -1$$

The solutions are $x = 4$ and $x = -1$.

2.3. QUADRATIC EQUATIONS

2.3.2 Completing the Square

Completing the square is another method for solving quadratic equations, particularly useful when the equation does not factor easily. The idea is to transform the quadratic equation into a perfect square trinomial, which can then be solved by taking the square root of both sides.

Example: Solve $x^2 + 6x + 5 = 0$ by completing the square

- **Step 1:** Move the constant term to the other side of the equation:
$$x^2 + 6x = -5$$

- **Step 2:** Add the square of half the coefficient of x to both sides:
$$x^2 + 6x + 9 = -5 + 9$$
$$x^2 + 6x + 9 = 4$$

- **Step 3:** Write the left side as a perfect square:
$$(x + 3)^2 = 4$$

- **Step 4:** Take the square root of both sides:
$$x + 3 = \pm 2$$

- **Step 5:** Solve for x:
$$x = -3 + 2 = -1 \quad \text{or} \quad x = -3 - 2 = -5$$

The solutions are $x = -1$ and $x = -5$.

2.3.3 The Quadratic Formula

If a quadratic equation cannot be factored easily, the quadratic formula can be used:

$$x = \frac{-b \pm \sqrt{b^2 - 4ac}}{2a}$$

Example: Solve $2x^2 + 3x - 2 = 0$ **using the quadratic formula**

- **Step 1:** Identify the coefficients: $a = 2, b = 3, c = -2$.
- **Step 2:** Substitute into the quadratic formula:

$$x = \frac{-3 \pm \sqrt{3^2 - 4(2)(-2)}}{2(2)}$$

- **Step 3:** Simplify the expression:

$$x = \frac{-3 \pm \sqrt{9 + 16}}{4}$$

$$x = \frac{-3 \pm \sqrt{25}}{4}$$

$$x = \frac{-3 \pm 5}{4}$$

- **Step 4:** Solve for the two possible values of x:

$$x = \frac{2}{4} = 0.5 \quad \text{or} \quad x = \frac{-8}{4} = -2$$

The solutions are $x = 0.5$ and $x = -2$.

2.3.4 Applications of Quadratic Equations

Quadratic equations frequently appear in various real-world contexts. Understanding how to apply them is essential for solving practical problems in fields such as physics, engineering, and economics.

Example: Projectile Motion

Consider the height h (in meters) of a ball thrown upwards with an initial velocity of 20 m/s from a height of 1.5 meters. The height of the ball at time t seconds is given by the quadratic equation:

$$h(t) = -4.9t^2 + 20t + 1.5$$

2.3. QUADRATIC EQUATIONS

To find out when the ball will hit the ground, set $h(t) = 0$:

$$-4.9t^2 + 20t + 1.5 = 0$$

This is a quadratic equation that can be solved using the quadratic formula:

$$t = \frac{-20 \pm \sqrt{20^2 - 4(-4.9)(1.5)}}{2(-4.9)}$$

Simplifying this will give the time t when the ball hits the ground.

Example: Maximizing Area

A farmer wants to create a rectangular pen with a fixed perimeter of 100 meters and wants to maximize the area. The area A is given by:

$$A = x(50 - x)$$

where x is the width. This quadratic equation can be solved by finding the vertex, which gives the maximum area.

Recap

In this chapter, we've explored the fundamental concepts of algebra, focusing on equations, inequalities, and polynomials. We've learned how to solve linear equations and systems of equations, how to factor polynomials, and how to solve quadratic equations using factoring, completing the square, and the quadratic formula. Additionally, we've seen how quadratic equations can be applied to real-world problems. These building blocks are essential as we continue to unlock more complex mathematical problems in the chapters to come.

Chapter 3

Exploring Functions

3.1 Introduction to Functions

Functions are one of the most important concepts in mathematics. A function represents a relationship between two sets of values, where each input value (from the domain) is associated with exactly one output value (from the range). Understanding functions is crucial for studying more advanced topics in mathematics, including calculus.

3.1.1 Definition and Notation of Functions

A function f from a set X to a set Y is a rule that assigns to each element x in X exactly one element y in Y. The notation $f : X \to Y$ is used to represent this relationship, where $y = f(x)$ is the output for a given input x.

Example: Definition of a Function

Consider the function $f(x) = 2x + 3$. This function takes an input x, multiplies it by 2, and then adds 3 to produce the output. For example, if $x = 4$:

$$f(4) = 2(4) + 3 = 8 + 3 = 11$$

Here, the input $x = 4$ is associated with the output $f(4) = 11$.

3.1.2 Domain and Range

The **domain** of a function is the set of all possible input values, while the **range** is the set of all possible output values. Determining the domain and range of a function is essential for understanding its behavior.

Example: Finding the Domain and Range

Consider the function $g(x) = \sqrt{x - 2}$.

- **Domain:** The expression inside the square root, $x - 2$, must be non-negative. Therefore, $x - 2 \geq 0$, or $x \geq 2$. The domain is $x \geq 2$.

- **Range:** Since the square root function produces non-negative outputs, the range of $g(x)$ is $g(x) \geq 0$.

3.1.3 Types of Functions

Functions can be classified into different types based on their behavior and the operations involved. Some common types of functions include linear, quadratic, polynomial, exponential, and logarithmic functions.

Example: Types of Functions

- **Linear Function:** $f(x) = 3x + 5$
- **Quadratic Function:** $f(x) = x^2 - 4x + 7$
- **Exponential Function:** $f(x) = 2^x$
- **Logarithmic Function:** $f(x) = \log(x)$

Each of these functions has distinct properties and graphs, which we will explore further in this chapter.

3.2 Linear and Quadratic Functions

Linear and quadratic functions are among the most fundamental types of functions in algebra. Understanding their properties and how to work with them is key to solving a wide range of mathematical problems.

3.2.1 Linear Functions

A linear function is a function of the form $f(x) = mx + b$, where m is the slope and b is the y-intercept. The graph of a linear function is a straight line.

Example: Graphing a Linear Function

Consider the function $f(x) = 2x - 3$. To graph this function, follow these steps:

- **Step 1:** Identify the slope $m = 2$ and the y-intercept $b = -3$.

- **Step 2:** Plot the y-intercept $(0, -3)$ on the graph.

- **Step 3:** Use the slope to find another point. Since the slope is 2, move up 2 units and right 1 unit from the y-intercept to find the point $(1, -1)$.

- **Step 4:** Draw a line through these points to complete the graph.

3.2. LINEAR AND QUADRATIC FUNCTIONS

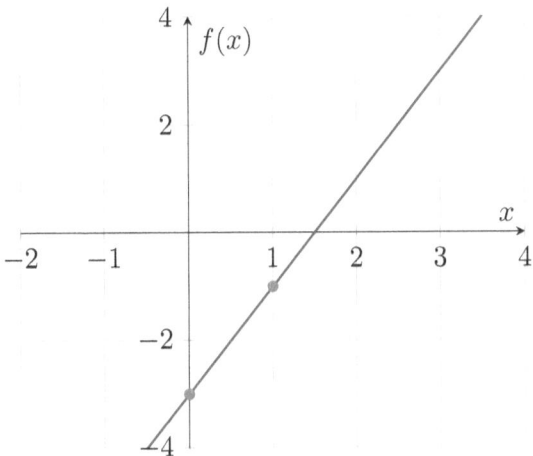

3.2.2 Quadratic Functions

A quadratic function is a function of the form $f(x) = ax^2 + bx + c$, where a, b, and c are constants. The graph of a quadratic function is a parabola, which opens upwards if $a > 0$ and downwards if $a < 0$.

Example: Graphing a Quadratic Function

Consider the function $f(x) = x^2 - 4x + 3$. To graph this function, follow these steps:

- **Step 1:** Identify the coefficients: $a = 1, b = -4, c = 3$.

- **Step 2:** Find the vertex using the formula $x = -\frac{b}{2a}$. For this function, $x = 2$.

- **Step 3:** Find the y-coordinate of the vertex by substituting $x = 2$ into the function: $f(2) = 2^2 - 4(2) + 3 = -1$. The vertex is $(2, -1)$.

- **Step 4:** Identify the y-intercept $(0, 3)$ and another point on the parabola, such as $(4, 3)$.

- **Step 5:** Plot these points and draw the parabola.

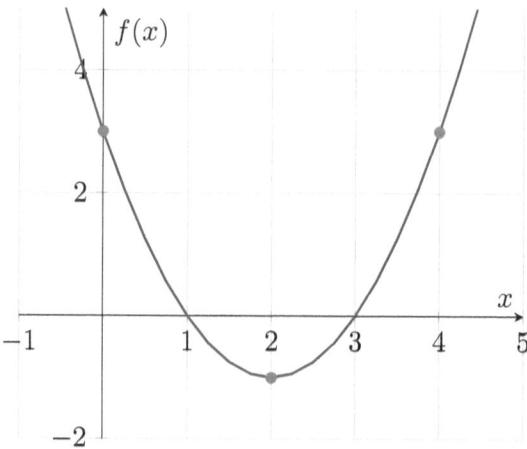

3.3 Advanced Functions

Beyond linear and quadratic functions, there are several other important types of functions, including exponential, logarithmic, rational, and radical functions. Each has unique properties and applications.

3.3.1 Exponential and Logarithmic Functions

Exponential functions are of the form $f(x) = a \cdot b^x$, where a and b are constants and $b > 0$. Logarithmic functions are the inverses of exponential functions, typically written as $f(x) = \log_b(x)$, where b is the base.

Example: Exponential Function

Consider the exponential function $f(x) = 3 \cdot 2^x$. This function grows rapidly as x increases because the base 2 is greater than 1. The graph of this function is a curve that increases exponentially.

3.3. ADVANCED FUNCTIONS

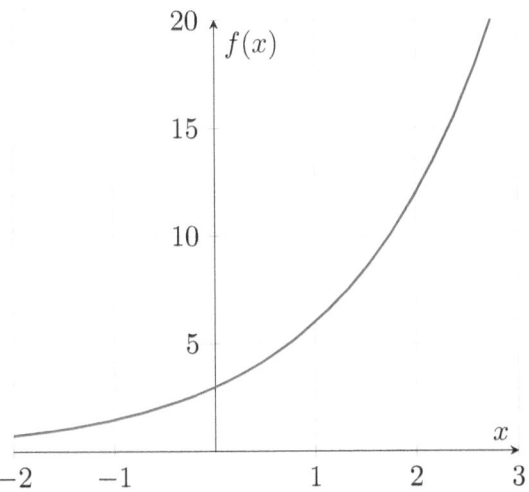

Example: Logarithmic Function

The logarithmic function $f(x) = \log_2(x)$ is the inverse of the exponential function $f(x) = 2^x$. The graph of $f(x) = \log_2(x)$ increases slowly and passes through the point $(1, 0)$.

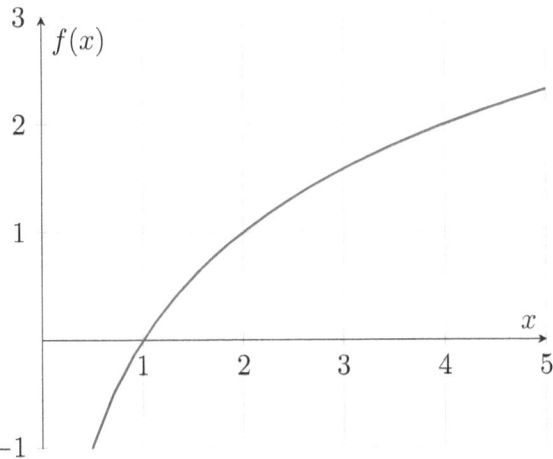

3.3.2 Rational and Radical Functions

Rational functions are functions of the form $f(x) = \frac{p(x)}{q(x)}$, where $p(x)$ and $q(x)$ are polynomials. Radical functions involve roots, such as square roots or cube roots.

Example: Rational Function

Consider the rational function $f(x) = \frac{1}{x-2}$. This function has a vertical asymptote at $x = 2$ because the denominator becomes zero at that point. The graph of this function approaches infinity as x approaches 2 from the left or right.

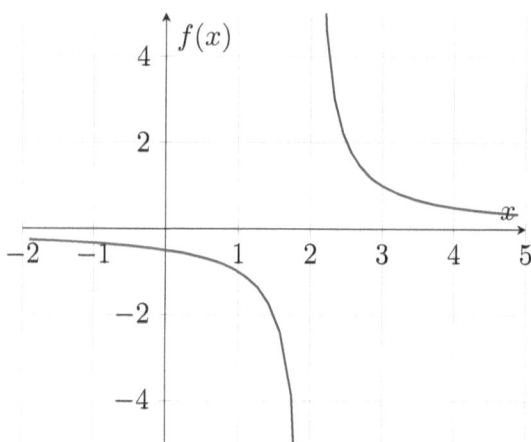

Example: Radical Function

The radical function $f(x) = \sqrt{x+3}$ is defined for $x \geq -3$. The graph of this function starts at $(-3, 0)$ and increases gradually, curving upwards.

3.3. ADVANCED FUNCTIONS

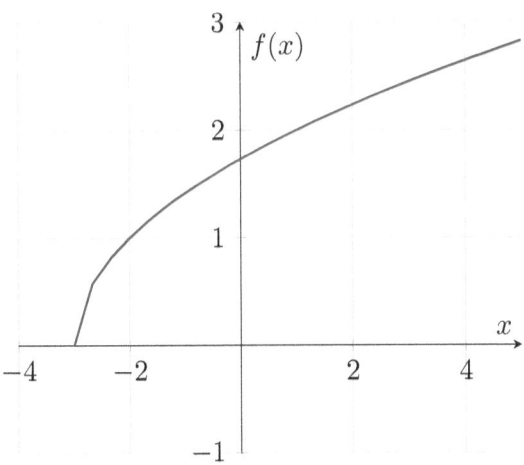

Recap

In this chapter, we've explored the concept of functions, starting with basic definitions and progressing through linear, quadratic, exponential, logarithmic, rational, and radical functions. Understanding these functions is essential for mastering more advanced mathematical topics. By learning how to identify, graph, and analyze these functions, you've gained the tools to solve a wide range of problems in algebra and beyond.

Chapter 4

Geometry Made Simple

4.1 Basic Geometric Concepts

Geometry is the branch of mathematics that deals with the properties and relationships of points, lines, angles, and shapes. Understanding these basic geometric concepts is essential for solving a wide range of problems in mathematics and the physical sciences.

4.1.1 Points, Lines, and Angles

In geometry, a **point** represents a location in space and has no size. A **line** is a straight path that extends infinitely in both directions, while a **line segment** has two endpoints and is finite in length. An **angle** is formed by two rays (or line segments) that share a common endpoint, called the **vertex**.

Example: Drawing Basic Geometric Figures

Here are basic geometric figures such as points, lines, and angles:

4.1. BASIC GEOMETRIC CONCEPTS

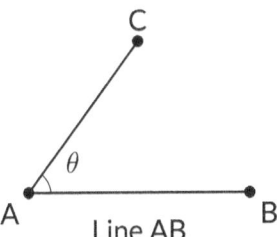

Line AB

4.1.2 Shapes and Their Properties

Shapes in geometry are defined by points connected by line segments. Common shapes include triangles, quadrilaterals, circles, and polygons. Each shape has unique properties that are crucial for understanding geometry.

Example: Understanding a Triangle

Consider a triangle with vertices A, B, and C. The sum of the interior angles of a triangle is always 180 degrees.

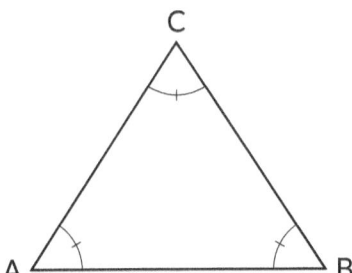

4.1.3 Properties of Triangles, Quadrilaterals, and Circles

Triangles are classified by their sides (equilateral, isosceles, scalene) and angles (acute, right, obtuse). Quadrilaterals include squares, rectangles, and parallelograms, each with specific properties regarding their sides and angles. Circles are defined by a center and radius, and important properties include the circumference and area.

Example: Properties of a Circle

A circle with center O and radius r has a circumference $C = 2\pi r$ and an area $A = \pi r^2$.

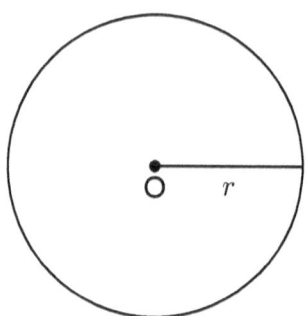

4.2 Perimeter, Area, and Volume

Calculating the perimeter, area, and volume of various shapes is a fundamental skill in geometry. The perimeter is the distance around a shape, the area is the amount of space it covers, and the volume is the space it occupies in three dimensions.

4.2.1 Calculating Perimeter and Area of Basic Shapes

For basic shapes, the perimeter and area can be calculated using standard formulas.

Example: Perimeter and Area of a Rectangle

For a rectangle with length l and width w, the perimeter P is given by:

$$P = 2l + 2w$$

The area A is:

$$A = lw$$

4.2. PERIMETER, AREA, AND VOLUME

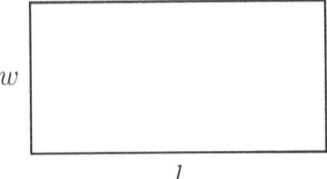

Example: Area of a Triangle

For a triangle with base b and height h, the area A is:

$$A = \frac{1}{2}bh$$

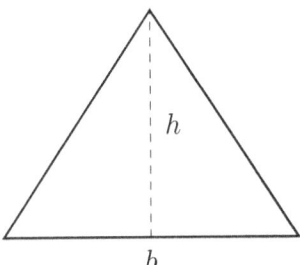

4.2.2 Volume of 3D Shapes

The volume of three-dimensional shapes like cubes, blocks, prisms, cylinders, pyramids, cones, and spheres can be calculated using specific formulas.

Example: Volume of a Cube

A cube is a special type of rectangular prism where all sides are equal in length. The volume V of a cube is given by:

$$V = s^3$$

where s is the length of a side.
Here is a 3D representation of a cube:

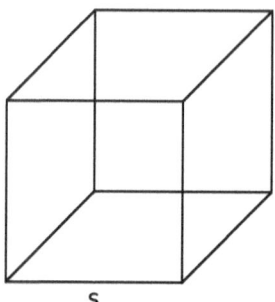

s

Example: Volume of a Rectangular Block

A rectangular block (or rectangular prism) is a 3D shape with rectangular faces. The volume V of a block is given by:

$$V = l \times w \times h$$

where l is the length, w is the width, and h is the height. Here is a 3D representation of a rectangular block:

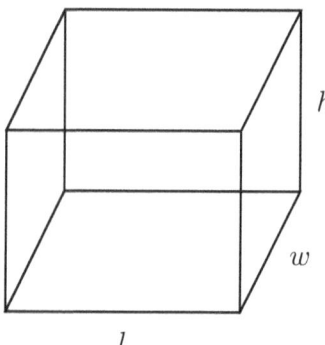

Example: Volume of a Prism

A prism is a 3D shape with two parallel and congruent bases. The volume V of a prism is given by:

$$V = \text{Base Area} \times \text{Height}$$

For a triangular prism with a triangular base of area A_b and height h:

4.2. PERIMETER, AREA, AND VOLUME

$$V = A_b \times h = \frac{1}{2} \times b \times h_b \times h$$

where b is the base length of the triangular face, h_b is the height of the triangular face, and h is the height of the prism.

Here is a 3D representation of a triangular prism:

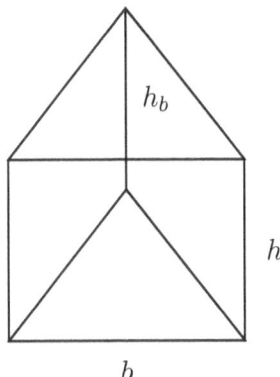

Example: Volume of a Cylinder

For a cylinder with radius r and height h, the volume V is:

$$V = \pi r^2 h$$

Here is a 3D representation of the cylinder:

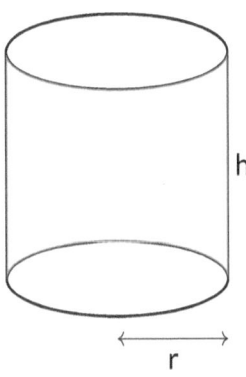

CHAPTER 4. GEOMETRY MADE SIMPLE

Example: Volume of a Pyramid

A pyramid is a 3D shape with a polygonal base and triangular faces that meet at a single point (the apex). The volume V of a pyramid is given by:

$$V = \frac{1}{3} \times \text{Base Area} \times \text{Height}$$

Here is a 3D representation of a pyramid:

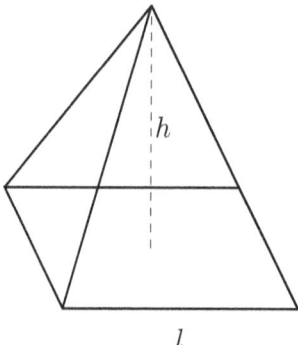

Example: Volume of a Cone

A cone is a 3D shape with a circular base and a curved surface that tapers to a point (the apex). The volume V of a cone is given by:

$$V = \frac{1}{3} \times \pi r^2 \times h$$

Here is a 3D representation of a cone:

4.3. COORDINATE GEOMETRY

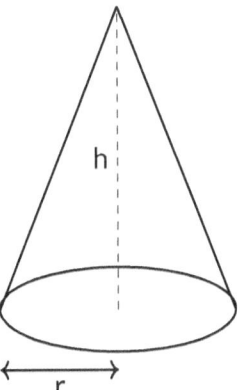

Example: Volume of a Sphere

A sphere is a perfectly round 3D shape, where every point on the surface is equidistant from the center. The volume V of a sphere is given by:

$$V = \frac{4}{3} \times \pi r^3$$

Here is a 3D representation of a sphere:

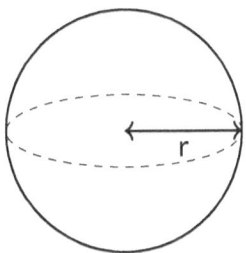

4.3 Coordinate Geometry

Coordinate geometry, also known as analytic geometry, is the study of geometry using a coordinate system. The Cartesian coordinate system is the most common, where each point is defined by an (x, y) pair.

4.3.1 Plotting Points and Lines

In the Cartesian plane, each point is represented by a pair of coordinates (x, y). A line can be represented by an equation of the form $y = mx + c$, where m is the slope and c is the y-intercept.

Example: Plotting Points and Drawing a Line

Consider the points $(1, 2)$ and $(3, 4)$. The line passing through these points can be drawn as follows:

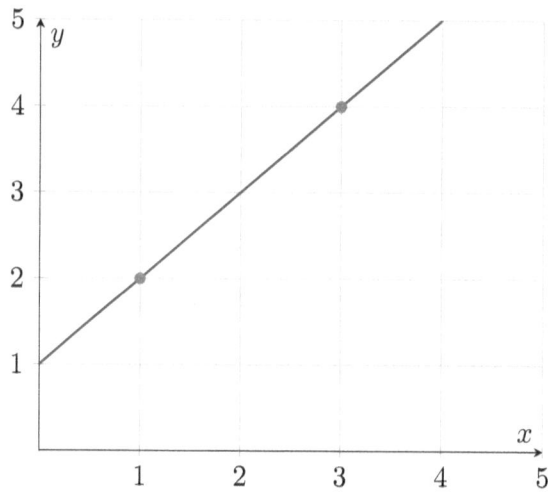

4.3.2 Distance and Midpoint Formulas

The distance between two points (x_1, y_1) and (x_2, y_2) in the Cartesian plane is given by:

$$d = \sqrt{(x_2 - x_1)^2 + (y_2 - y_1)^2}$$

The midpoint M of the segment connecting these two points is:

$$M = \left(\frac{x_1 + x_2}{2}, \frac{y_1 + y_2}{2} \right)$$

4.3. COORDINATE GEOMETRY

Example: Finding the Distance and Midpoint

For points $A(1, 2)$ and $B(4, 6)$:

$$d = \sqrt{(4-1)^2 + (6-2)^2} = \sqrt{9 + 16} = \sqrt{25} = 5$$

$$M = \left(\frac{1+4}{2}, \frac{2+6}{2}\right) = \left(\frac{5}{2}, \frac{8}{2}\right) = (2.5, 4)$$

Recap

In this chapter, we've explored the basic concepts of geometry, including points, lines, angles, and shapes. We've also covered how to calculate the perimeter, area, and volume of various geometric figures, as well as the basics of coordinate geometry. These foundational skills are essential for solving more complex problems in mathematics and the physical sciences.

Chapter 5

Trigonometry Without Tears

Trigonometry is a powerful mathematical tool that helps us understand the relationships between angles and sides of triangles. It's not just about solving triangles; trigonometry is essential in fields like physics, engineering, and even music theory. In this chapter, we'll explore trigonometric ratios, the unit circle, and how to use trigonometric identities and equations to solve problems with ease.

5.1 Trigonometric Ratios

Trigonometric ratios are the foundation of trigonometry, helping us relate the angles of a triangle to the lengths of its sides. These ratios—sine, cosine, and tangent—are particularly useful in right-angled triangles.

5.1.1 Understanding Sine, Cosine, and Tangent

The three primary trigonometric ratios are sine (sin), cosine (cos), and tangent (tan). These ratios are defined as follows in a right-angled triangle:

- $\sin \theta = \frac{\text{Opposite}}{\text{Hypotenuse}} = \frac{a}{c}$
- $\cos \theta = \frac{\text{Adjacent}}{\text{Hypotenuse}} = \frac{b}{c}$

5.2. THE UNIT CIRCLE

- $\tan\theta = \frac{\text{Opposite}}{\text{Adjacent}} = \frac{a}{b}$

Example: Understanding Trigonometric Ratios

Consider a right-angled triangle with angle θ, where $a = 3$, $b = 4$, and $c = 5$:

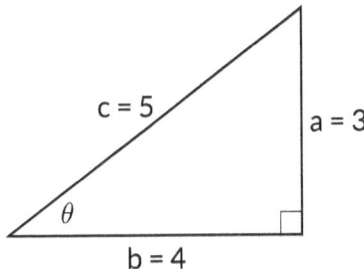

Using the definitions of sine, cosine, and tangent, we calculate:

$$\sin\theta = \frac{3}{5}, \quad \cos\theta = \frac{4}{5}, \quad \tan\theta = \frac{3}{4}$$

5.1.2 Applications of Trigonometric Ratios in Right Triangles

Trigonometric ratios are essential in solving problems involving right triangles, such as finding unknown sides or angles.

Example: Finding an Unknown Side

Given a right triangle where $a = 5$ and $\theta = 30°$, find the length of the hypotenuse c.

$$\sin 30° = \frac{a}{c} = \frac{5}{c} \quad \Rightarrow \quad c = \frac{5}{\sin 30°} = \frac{5}{0.5} = 10$$

5.2 The Unit Circle

The unit circle is a fundamental concept in trigonometry, allowing us to extend trigonometric functions beyond right triangles to all angles.

5.2.1 The Unit Circle and Its Significance in Trigonometry

The unit circle is a circle with a radius of 1 centered at the origin of a coordinate plane. It allows us to define trigonometric functions for any angle.

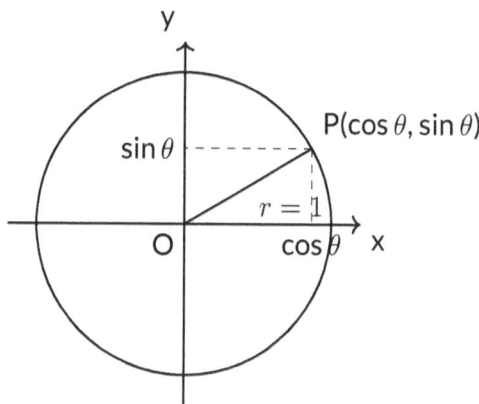

5.2.2 Using the Unit Circle to Find Trigonometric Values

The unit circle helps us determine the sine, cosine, and tangent of angles beyond those in right triangles. For any angle θ, the coordinates of the corresponding point on the unit circle are $(\cos\theta, \sin\theta)$.

Example: Using the Unit Circle

To find $\sin 45°$ and $\cos 45°$, locate the point on the unit circle at $45°$:

$$\sin 45° = \cos 45° = \frac{\sqrt{2}}{2}$$

5.3 Trigonometric Identities and Equations

Trigonometric identities are equations involving trigonometric functions that hold true for all values of the variable. These identities simplify complex expressions and are key to solving trigonometric equations.

5.3. TRIGONOMETRIC IDENTITIES AND EQUATIONS

5.3.1 Fundamental Identities and How to Use Them

Some of the most important trigonometric identities include:

- Pythagorean Identity: $\sin^2\theta + \cos^2\theta = 1$
- Tangent Identity: $\tan\theta = \frac{\sin\theta}{\cos\theta}$
- Reciprocal Identities:

$$\csc\theta = \frac{1}{\sin\theta}, \quad \sec\theta = \frac{1}{\cos\theta}, \quad \cot\theta = \frac{1}{\tan\theta}$$

Example: Using the Pythagorean Identity

Given $\sin\theta = \frac{3}{5}$, find $\cos\theta$ using the Pythagorean identity.

$$\sin^2\theta + \cos^2\theta = 1$$

$$\left(\frac{3}{5}\right)^2 + \cos^2\theta = 1$$

$$\frac{9}{25} + \cos^2\theta = 1$$

$$\cos^2\theta = \frac{16}{25} \quad \Rightarrow \quad \cos\theta = \frac{4}{5}$$

5.3.2 Solving Basic Trigonometric Equations

Trigonometric equations involve trigonometric functions and can often be solved using identities.

Example: Solving a Trigonometric Equation

Solve the equation $2\sin x - 1 = 0$ for x in the interval $[0, 2\pi]$.

$$2\sin x = 1 \quad \Rightarrow \quad \sin x = \frac{1}{2}$$

$$x = \frac{\pi}{6}, \quad \text{or} \quad x = \frac{5\pi}{6}$$

Recap

In this chapter, we've explored trigonometric ratios, the unit circle, and essential trigonometric identities and equations. We've also seen how these concepts apply to solving right triangles and determining trigonometric values using the unit circle. These tools form the foundation of trigonometry and are crucial for analyzing various mathematical and real-world problems involving angles and distances.

Chapter 6

Probability and Statistics

Probability and statistics form the cornerstone of decision-making in fields like science, business, and economics. They help us understand randomness, make predictions, and interpret data. In this chapter, we'll explore the basics of probability, how to describe data using descriptive statistics, and learn the foundations of data interpretation and analysis.

6.1 Basics of Probability

Probability measures the likelihood of an event happening, ranging from 0 (impossible) to 1 (certain). It provides the framework for making informed predictions about outcomes.

6.1.1 Probability Concepts and Rules

The probability of an event A, denoted $P(A)$, is calculated by dividing the number of favorable outcomes by the total number of possible outcomes.

$$P(A) = \frac{\text{Number of favorable outcomes}}{\text{Total number of outcomes}}$$

For example, the probability of rolling a 3 on a fair 6-sided die is:

$$P(3) = \frac{1}{6}$$

Example: Probability of Rolling a Die

If we roll a 6-sided die, the possible outcomes are $\{1, 2, 3, 4, 5, 6\}$. The probability of rolling an even number is:

$$P(\text{even number}) = \frac{3 \text{ even outcomes}}{6 \text{ total outcomes}} = \frac{3}{6} = 0.5$$

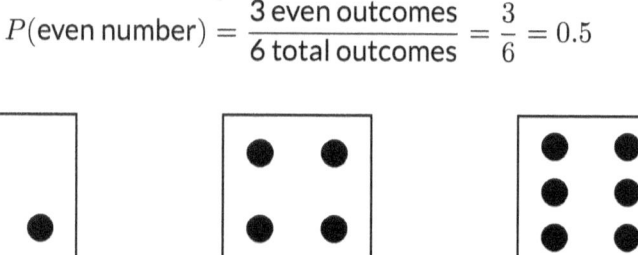

Rolling a "2"　　　　Rolling a "4"　　　　Rolling a "6"

6.1.2 Simple and Compound Events

Events can be classified as simple or compound. A **simple event** is a single outcome, while a **compound event** involves two or more simple events.

For example, rolling a 4 on a die is a simple event, while rolling a 4 and flipping heads on a coin toss is a compound event.

Example: Compound Event

If we roll a die and flip a coin, what's the probability of rolling a 3 and getting heads?

$$P(\text{rolling 3 and heads}) = P(\text{rolling 3}) \times P(\text{heads}) = \frac{1}{6} \times \frac{1}{2} = \frac{1}{12}$$

6.2 Descriptive Statistics

Descriptive statistics helps us summarize and describe the essential features of data. We'll explore two key aspects: measures of central tendency and measures of spread.

6.2.1 Measures of Central Tendency

The **mean**, **median**, and **mode** are the most common measures of central tendency, which give us an idea of the "center" of a data set.

Mean

The mean is the average of a data set, calculated by adding all the values together and dividing by the number of values.

$$\text{Mean} = \frac{\sum \text{Data Points}}{\text{Number of Data Points}}$$

Example: Calculating the Mean

Consider the following data set: $\{2, 4, 6, 8, 10\}$. The mean is:

$$\text{Mean} = \frac{2+4+6+8+10}{5} = \frac{30}{5} = 6$$

Median

The median is the middle value of a data set when the data is arranged in ascending order. If there's an even number of data points, the median is the average of the two middle values.

Example: Finding the Median

For the data set $\{1, 3, 7, 9, 12\}$, the median is 7 because it is the middle value. If the data set were $\{1, 3, 7, 9\}$, the median would be the average of 3 and 7, which is $\frac{3+7}{2} = 5$.

Mode

The mode is the value that appears most frequently in a data set. Some data sets may have more than one mode, while others may not have any.

Example: Identifying the Mode

In the data set $\{4, 4, 2, 5, 2, 4, 6\}$, the mode is 4 because it appears the most often.

6.2.2 Measures of Spread

The **range**, **variance**, and **standard deviation** help us understand the variability or spread of data.

Range

The range is the difference between the highest and lowest values in a data set.

$$\text{Range} = \text{Maximum Value} - \text{Minimum Value}$$

Example: Calculating the Range

For the data set $\{2, 4, 6, 8, 10\}$, the range is:

$$\text{Range} = 10 - 2 = 8$$

Variance and Standard Deviation

Variance and standard deviation give us more detailed information about how much the data points differ from the mean.

- **Variance** is the average of the squared differences from the mean.
- **Standard Deviation** is the square root of the variance, giving us a measure of spread in the same units as the data.

6.3. DATA INTERPRETATION AND ANALYSIS

$$\text{Variance} = \frac{\sum (x_i - \bar{x})^2}{n}$$

$$\text{Standard Deviation} = \sqrt{\text{Variance}}$$

6.3 Data Interpretation and Analysis

Data interpretation is about making sense of data through the use of graphs, charts, and statistical tools. It is an essential skill in both academic and real-world settings.

6.3.1 Reading and Interpreting Graphs and Charts

Graphs and charts provide a visual representation of data, helping us identify patterns, trends, and outliers. Common types include bar charts, pie charts, and line graphs.

Example: Interpreting a Bar Chart

The following bar chart shows the number of books read by students in a class:

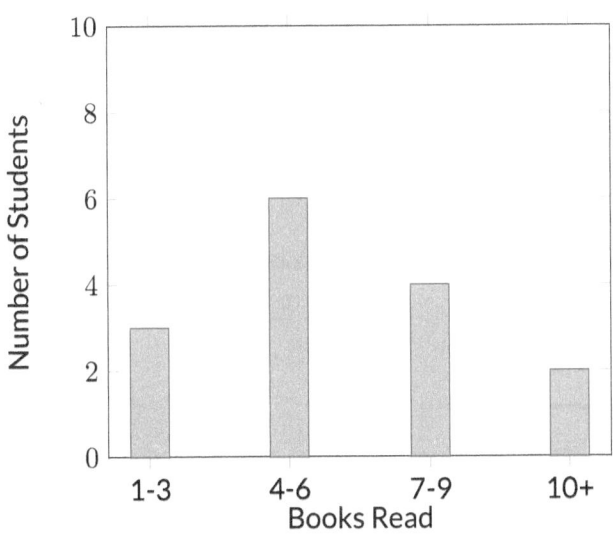

From this bar chart, we can see that most students read between 4 and 6 books.

6.3.2 Basics of Statistical Inference

Statistical inference involves making predictions or generalizations about a population based on a sample of data. This process helps us make educated guesses about data we haven't yet observed.
There are two main types of statistical inference:

- **Estimation**: Using sample data to estimate population parameters.

- **Hypothesis Testing**: Making decisions about the population based on sample data.

Example: Hypothesis Testing

A researcher claims that the average height of students in a school is 160 cm. To test this, we collect a sample of 30 students, and their average height is found to be 158 cm with a standard deviation of 5 cm. We will perform a hypothesis test at a significance level of $\alpha = 0.05$.

- **Step 1: State the Hypotheses**

    We begin by stating the null hypothesis (H_0) and the alternative hypothesis (H_1):

    - H_0: The average height of students in the school is 160 cm. (The claim is true.)

    $$H_0 : \mu = 160$$

    - H_1: The average height of students in the school is different from 160 cm. (The claim is not true.)

    $$H_1 : \mu \neq 160$$

    This is a two-tailed test because we are testing whether the average height is different from 160 cm, either higher or lower.

6.3. DATA INTERPRETATION AND ANALYSIS

- **Step 2: Set the Significance Level**

  We will use a significance level (α) of 0.05. This means that we are willing to accept a 5% chance of rejecting the null hypothesis when it is true (Type I error).

  $$\alpha = 0.05$$

- **Step 3: Compute the Test Statistic**

  The test statistic for a one-sample t-test is calculated using the formula:

  $$t = \frac{\bar{x} - \mu_0}{\frac{s}{\sqrt{n}}}$$

  Where:

  - $\bar{x} = 158$ cm is the sample mean,
  - $\mu_0 = 160$ cm is the population mean under the null hypothesis,
  - $s = 5$ cm is the sample standard deviation, and
  - $n = 30$ is the sample size.

  Substituting the values into the formula:

  $$t = \frac{158 - 160}{\frac{5}{\sqrt{30}}} = \frac{-2}{\frac{5}{\sqrt{30}}} = \frac{-2}{0.913} \approx -2.19$$

  The calculated test statistic is $t \approx -2.19$.

- **Step 4: Determine the Critical Value or P-value**

  Since we are conducting a two-tailed test at $\alpha = 0.05$ with $n - 1 = 30 - 1 = 29$ degrees of freedom, we can look up the critical value for t from a t-distribution table.

  For a two-tailed test with $\alpha = 0.05$ and 29 degrees of freedom, the critical values are:

$$t_{\text{critical}} = \pm 2.045$$

Alternatively, we can compute the p-value corresponding to the test statistic $t = -2.19$. Using a statistical software or a t-distribution calculator, we find:

$$\text{p-value} \approx 0.036$$

- **Step 5: Make a Decision and Interpret the Result**

    To decide whether to reject or fail to reject the null hypothesis, we compare the p-value to the significance level α.

    Since p-value $= 0.036$ is less than $\alpha = 0.05$, we reject the null hypothesis.

    Alternatively, if using the critical value approach:

    The calculated test statistic $t = -2.19$ falls outside the range $-2.045 \leq t \leq 2.045$, so we reject the null hypothesis.

- **Conclusion**

    There is sufficient evidence at $\alpha = 0.05$ to conclude that the average height of students in the school is significantly different from 160 cm. Therefore, the researcher's claim is not supported by the sample data.

Recap

In this chapter, we covered the basics of probability, including simple and compound events. We also explored descriptive statistics, learning how to measure central tendency and spread in a data set. Finally, we introduced the key concepts of data interpretation, including reading graphs and understanding the basics of statistical inference. Mastery of these topics is crucial for understanding data in both everyday situations and more complex scientific studies.

Chapter 7

Advanced Algebra and Pre-Calculus

Advanced algebra and pre-calculus form the bridge between high school algebra and calculus. In this chapter, we will explore essential topics such as sequences and series, matrices and determinants, and an introduction to limits. Mastering these concepts prepares you for more complex topics in calculus and higher mathematics.

7.1 Sequences and Series

Sequences and series are essential components of advanced algebra, used extensively in calculus, number theory, and other mathematical areas.

7.1.1 Arithmetic and Geometric Sequences

A **sequence** is a list of numbers following a specific pattern. In an **arithmetic sequence**, the difference between consecutive terms is constant, while in a **geometric sequence**, each term is obtained by multiplying the previous term by a constant.

Arithmetic Sequence

The general form of an arithmetic sequence is given by:

$$a_n = a_1 + (n-1)d$$

Where:

- a_n is the n-th term,
- a_1 is the first term,
- d is the common difference between terms, and
- n is the position of the term in the sequence.

Example: Arithmetic Sequence

Consider the arithmetic sequence $3, 7, 11, 15, \ldots$, where $a_1 = 3$ and $d = 4$. The 5-th term is:

$$a_5 = 3 + (5-1) \times 4 = 3 + 16 = 19$$

Geometric Sequence

In a geometric sequence, the ratio between consecutive terms is constant. The general form of a geometric sequence is:

$$a_n = a_1 \cdot r^{n-1}$$

Where:

- a_n is the n-th term,
- a_1 is the first term,
- r is the common ratio between terms, and
- n is the position of the term in the sequence.

7.1. SEQUENCES AND SERIES

Example: Geometric Sequence

Consider the geometric sequence $2, 6, 18, 54, \ldots$, where $a_1 = 2$ and $r = 3$. The 4-th term is:

$$a_4 = 2 \cdot 3^{4-1} = 2 \cdot 27 = 54$$

7.1.2 Summation Notation and Series

A **series** is the sum of the terms in a sequence. Summation notation, represented by the sigma (Σ) symbol, is used to describe the sum of a sequence's terms.

Arithmetic Series

The sum of the first n terms of an arithmetic series is given by the formula:

$$S_n = \frac{n}{2}(2a_1 + (n-1)d)$$

Example: Arithmetic Series

The sum of the first 5 terms of the arithmetic sequence $3, 7, 11, 15, \ldots$ is:

$$S_5 = \frac{5}{2}(2(3) + (5-1)(4)) = \frac{5}{2}(6+16) = \frac{5}{2} \cdot 22 = 55$$

Geometric Series

The sum of the first n terms of a geometric series is given by the formula:

$$S_n = a_1 \cdot \frac{1 - r^n}{1 - r} \quad \text{for} \quad r \neq 1$$

Example: Geometric Series

The sum of the first 4 terms of the geometric sequence $2, 6, 18, 54, \ldots$ is:

$$S_4 = 2 \cdot \frac{1 - 3^4}{1 - 3} = 2 \cdot \frac{1 - 81}{-2} = 2 \cdot 40 = 80$$

7.2 Matrices and Determinants

Matrices are powerful tools used in algebra, geometry, and computer science to organize and manipulate data. Determinants provide useful information about matrices, including whether a matrix has an inverse.

7.2.1 Introduction to Matrices and Operations

A **matrix** is a rectangular array of numbers arranged in rows and columns. Matrices are often used to solve systems of equations, represent transformations, and model data.

Matrix Addition and Subtraction

Matrices can be added or subtracted if they have the same dimensions. The operation is performed element-wise.

Example: Matrix Addition

Consider two matrices:

$$A = \begin{bmatrix} 1 & 2 \\ 3 & 4 \end{bmatrix}, \quad B = \begin{bmatrix} 5 & 6 \\ 7 & 8 \end{bmatrix}$$

The sum of A and B is:

$$A + B = \begin{bmatrix} 1+5 & 2+6 \\ 3+7 & 4+8 \end{bmatrix} = \begin{bmatrix} 6 & 8 \\ 10 & 12 \end{bmatrix}$$

7.2. MATRICES AND DETERMINANTS

Matrix Multiplication

Matrix multiplication is not element-wise. The product of two matrices A and B is defined if the number of columns in A matches the number of rows in B.

Example: Matrix Multiplication

Let $A = \begin{bmatrix} 1 & 2 \\ 3 & 4 \end{bmatrix}$ and $B = \begin{bmatrix} 2 & 0 \\ 1 & 3 \end{bmatrix}$. The product AB is:

$$AB = \begin{bmatrix} 1 \cdot 2 + 2 \cdot 1 & 1 \cdot 0 + 2 \cdot 3 \\ 3 \cdot 2 + 4 \cdot 1 & 3 \cdot 0 + 4 \cdot 3 \end{bmatrix} = \begin{bmatrix} 4 & 6 \\ 10 & 12 \end{bmatrix}$$

7.2.2 Determinants and Their Applications

The **determinant** of a square matrix provides information about the matrix's properties. For a 2×2 matrix A, the determinant is calculated as:

$$\det(A) = \begin{vmatrix} a & b \\ c & d \end{vmatrix} = ad - bc$$

Example: Determinant of a 2×2 Matrix

Let $A = \begin{bmatrix} 3 & 2 \\ 1 & 4 \end{bmatrix}$. The determinant of A is:

$$\det(A) = 3 \times 4 - 2 \times 1 = 12 - 2 = 10$$

The determinant can also be used to determine if a matrix has an inverse. A matrix is invertible if and only if its determinant is non-zero.

7.2.3 Applications of Matrices

Solving Systems of Equations Using Matrices

One of the most powerful applications of matrices is solving systems of linear equations. By using matrices, we can express a system of equations in compact form and solve it using methods like Gaussian elimination or matrix inversion.

A system of linear equations can be written in matrix form as:

$$AX = B$$

Where:

- A is the coefficient matrix,
- X is the column vector of variables, and
- B is the column vector of constants.

Example: Solving a System Using Matrices

Consider the system of equations:

$$x + 2y = 5$$
$$3x - y = 4$$

We can express this system in matrix form as:

$$\begin{bmatrix} 1 & 2 \\ 3 & -1 \end{bmatrix} \begin{bmatrix} x \\ y \end{bmatrix} = \begin{bmatrix} 5 \\ 4 \end{bmatrix}$$

We can solve this system by finding the inverse of matrix A (if it exists) and multiplying both sides of the equation by A^{-1}:

$$X = A^{-1}B$$

First, we compute the determinant of A:

$$\det(A) = \begin{vmatrix} 1 & 2 \\ 3 & -1 \end{vmatrix} = 1(-1) - 2(3) = -1 - 6 = -7$$

Since the determinant is non-zero, the matrix A is invertible. Now, we can find A^{-1} and solve the system:

$$A^{-1} = \frac{1}{\det(A)} \begin{bmatrix} -1 & -2 \\ -3 & 1 \end{bmatrix} = \frac{1}{-7} \begin{bmatrix} -1 & -2 \\ -3 & 1 \end{bmatrix}$$

7.2. MATRICES AND DETERMINANTS

Multiplying A^{-1} by B gives us the solution for X:

$$X = \frac{1}{-7}\begin{bmatrix} -1 & -2 \\ -3 & 1 \end{bmatrix}\begin{bmatrix} 5 \\ 4 \end{bmatrix} = \frac{1}{-7}\begin{bmatrix} -5-8 \\ -15+4 \end{bmatrix} = \frac{1}{-7}\begin{bmatrix} -13 \\ -11 \end{bmatrix} = \begin{bmatrix} 13/7 \\ 11/7 \end{bmatrix}$$

Thus, the solution to the system is $x = \frac{13}{7}$ and $y = \frac{11}{7}$.

Representing Transformations Using Matrices

Matrices are also used extensively to represent geometric transformations. Common transformations include rotations, reflections, and scaling in two or three dimensions. These transformations can be applied to vectors and points in space by multiplying the transformation matrix by the point or vector.

2D Transformations

In two dimensions, transformations are represented by 2×2 matrices. For example:

- A rotation by an angle θ is represented by the matrix:

$$R = \begin{bmatrix} \cos\theta & -\sin\theta \\ \sin\theta & \cos\theta \end{bmatrix}$$

- A scaling transformation, which stretches or shrinks a figure by a factor k, is represented by the matrix:

$$S = \begin{bmatrix} k & 0 \\ 0 & k \end{bmatrix}$$

- A reflection across the x-axis is represented by the matrix:

$$M = \begin{bmatrix} 1 & 0 \\ 0 & -1 \end{bmatrix}$$

Example: Rotation Using Matrices

Consider the point $P(1, 0)$ in two-dimensional space. To rotate this point by 90 degrees counterclockwise, we use the rotation matrix with $\theta = 90°$ (or $\theta = \frac{\pi}{2}$):

$$R = \begin{bmatrix} \cos\frac{\pi}{2} & -\sin\frac{\pi}{2} \\ \sin\frac{\pi}{2} & \cos\frac{\pi}{2} \end{bmatrix} = \begin{bmatrix} 0 & -1 \\ 1 & 0 \end{bmatrix}$$

Multiplying R by the coordinates of P:

$$R \begin{bmatrix} 1 \\ 0 \end{bmatrix} = \begin{bmatrix} 0 & -1 \\ 1 & 0 \end{bmatrix} \begin{bmatrix} 1 \\ 0 \end{bmatrix} = \begin{bmatrix} 0 \\ 1 \end{bmatrix}$$

The new coordinates of the point after rotation are $(0, 1)$.

7.3 Introduction to Limits

The concept of limits is fundamental in calculus, used to define derivatives, integrals, and continuity. A limit helps us understand the behavior of a function as the input approaches a certain value.

7.3.1 Concept of Limits and Their Importance in Calculus

The limit of a function $f(x)$ as x approaches a value c is written as:

$$\lim_{x \to c} f(x)$$

This represents the value that $f(x)$ gets closer to as x approaches c.

Example: Simple Limit

Consider the function $f(x) = 2x + 3$. The limit as x approaches 2 is:

$$\lim_{x \to 2}(2x + 3) = 2(2) + 3 = 4 + 3 = 7$$

7.3. INTRODUCTION TO LIMITS

7.3.2 Evaluating Limits Using Algebraic Techniques

Some limits can be evaluated directly by substituting the value of x into the function, while others require algebraic manipulation.

Example: Limit Requiring Algebraic Manipulation

Consider the limit:

$$\lim_{x \to 1} \frac{x^2 - 1}{x - 1}$$

Direct substitution of $x = 1$ leads to $\frac{0}{0}$, an indeterminate form. However, we can factor the numerator:

$$\lim_{x \to 1} \frac{(x - 1)(x + 1)}{x - 1}$$

Canceling the $(x - 1)$ terms:

$$\lim_{x \to 1} (x + 1) = 1 + 1 = 2$$

Conclusion on Limits

Limits are a powerful tool in understanding the behavior of functions near specific points. As we move toward calculus, limits will form the basis for concepts like differentiation and integration.

Recap

In this chapter, we explored the basics of sequences and series, including arithmetic and geometric sequences. We also introduced matrices and determinants, key tools for solving systems of equations and understanding matrix properties. We added practical insights into solving systems of equations and representing transformations using matrices. Finally, we laid the groundwork for calculus by introducing the concept of limits and techniques for evaluating them. These topics are fundamental to mastering advanced mathematics and preparing for calculus.

Chapter 8

Introduction to Calculus

Calculus is the study of how things change. It's a powerful tool used in many fields such as physics, engineering, economics, and biology. The two main branches of calculus are differentiation and integration, which deal with rates of change and accumulation, respectively. In this chapter, we will introduce these concepts and explore their applications.

8.1 The Concept of the Derivative

The derivative measures how a function changes as its input changes. It is commonly understood as the rate of change or the slope of a curve at a given point.

8.1.1 Understanding Derivatives as Rates of Change

The derivative of a function $f(x)$ at a point $x = a$ is defined as the limit:

$$f'(a) = \lim_{h \to 0} \frac{f(a+h) - f(a)}{h}$$

This formula represents the instantaneous rate of change of $f(x)$ with respect to x at $x = a$.

8.1. THE CONCEPT OF THE DERIVATIVE

Example: Derivative of a Linear Function

Consider the function $f(x) = 3x + 2$. The derivative $f'(x)$ measures the rate of change of $f(x)$ with respect to x. Using the definition of the derivative:

$$f'(x) = \lim_{h \to 0} \frac{(3(x+h) + 2) - (3x + 2)}{h} = \lim_{h \to 0} \frac{3h}{h} = 3$$

Thus, the derivative of $f(x) = 3x + 2$ is $f'(x) = 3$, meaning the rate of change is constant.

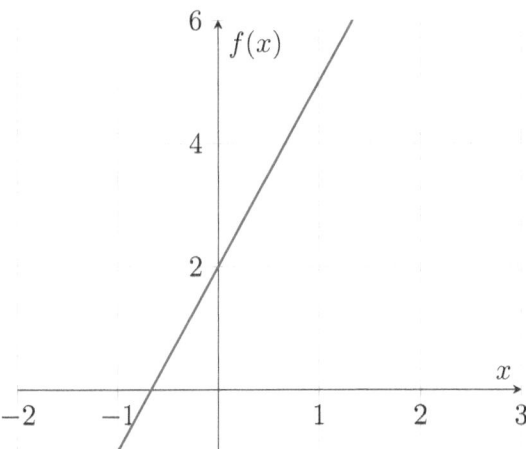

Figure 8.1: Graph of $f(x) = 3x + 2$ showing the constant slope.

8.1.2 Basic Differentiation Rules

To calculate derivatives efficiently, we use differentiation rules. Here are some of the most common ones:

- **Power Rule:** If $f(x) = x^n$, then $f'(x) = nx^{n-1}$.

- **Constant Rule:** The derivative of a constant is zero. If $f(x) = c$, then $f'(x) = 0$.

- **Sum Rule:** The derivative of the sum of two functions is the sum of their derivatives. If $f(x) = g(x) + h(x)$, then $f'(x) = g'(x) + h'(x)$.

- **Product Rule:** If $f(x) = g(x)h(x)$, then $f'(x) = g'(x)h(x) + g(x)h'(x)$.

- **Quotient Rule:** If $f(x) = \frac{g(x)}{h(x)}$, then $f'(x) = \frac{g'(x)h(x) - g(x)h'(x)}{[h(x)]^2}$.

Example: Using the Power Rule

Find the derivative of $f(x) = x^4$:

$$f'(x) = 4x^3$$

8.2 Applications of Derivatives

Derivatives have a wide range of applications in real-world scenarios, from calculating velocity and acceleration to optimizing functions and understanding how rates of change interact.

8.2.1 Derivatives in Motion, Optimization, and Curve Sketching

The derivative plays an important role in describing motion, such as finding the velocity or acceleration of an object. It is also used to optimize functions and find critical points (maxima and minima).

Example: Velocity and Acceleration

If the position of an object is given by $s(t) = 5t^2$, its velocity is the derivative of position with respect to time, and acceleration is the derivative of velocity.

$$v(t) = \frac{ds}{dt} = 10t$$

$$a(t) = \frac{dv}{dt} = 10$$

The velocity is $10t$ and the acceleration is constant at 10 units per time squared.

8.2.2 Related Rates and Real-World Applications

Related rates refer to problems where two or more quantities change with respect to time and are related to each other through an equation. These types of problems often arise in physics, engineering, and other fields.

Example: Related Rates Problem

A balloon is being inflated, and its radius is increasing at a rate of 3 cm/min. How fast is the volume of the balloon increasing when the radius is 10 cm?

We start with the formula for the volume of a sphere:

$$V = \frac{4}{3}\pi r^3$$

Taking the derivative of both sides with respect to time t:

$$\frac{dV}{dt} = 4\pi r^2 \frac{dr}{dt}$$

Substituting $r = 10$ cm and $\frac{dr}{dt} = 3$ cm/min:

$$\frac{dV}{dt} = 4\pi (10)^2 \times 3 = 1200\pi \text{ cm}^3/\text{min}$$

Thus, the volume is increasing at a rate of 1200π cm^3/min.

8.3 The Integral: Area and Accumulation

While derivatives measure rates of change, integrals measure accumulation. The integral allows us to calculate areas under curves and solve problems involving accumulation, such as finding the total distance traveled by an object or the area under a curve.

8.3.1 Basic Integration Techniques

The integral is the reverse operation of the derivative. The indefinite integral (or antiderivative) of a function $f(x)$ is a function $F(x)$ such that $F'(x) = f(x)$. The process of finding an integral is called integration.

The basic integral rules include:

- **Power Rule:** $\int x^n \, dx = \frac{x^{n+1}}{n+1} + C$ where $n \neq -1$
- **Constant Rule:** $\int c \, dx = cx + C$
- **Sum Rule:** $\int [f(x) + g(x)] \, dx = \int f(x) \, dx + \int g(x) \, dx$

Example: Using the Power Rule for Integration

Find the integral of $f(x) = x^3$:

$$\int x^3 \, dx = \frac{x^4}{4} + C$$

8.3.2 Applications of Integrals in Calculating Areas and Solving Problems of Accumulation

Calculating Areas

One of the most important applications of integrals is calculating the area under a curve. If $f(x)$ is a positive function, then the definite integral of $f(x)$ from a to b gives the area under the curve between $x = a$ and $x = b$:

$$\int_a^b f(x) \, dx$$

Example: Calculating the Area Under a Curve

Find the area under the curve $f(x) = x^2$ from $x = 0$ to $x = 2$:

$$\int_0^2 x^2 \, dx = \left[\frac{x^3}{3}\right]_0^2 = \frac{8}{3}$$

Thus, the area under the curve is $\frac{8}{3}$.

8.3. THE INTEGRAL: AREA AND ACCUMULATION

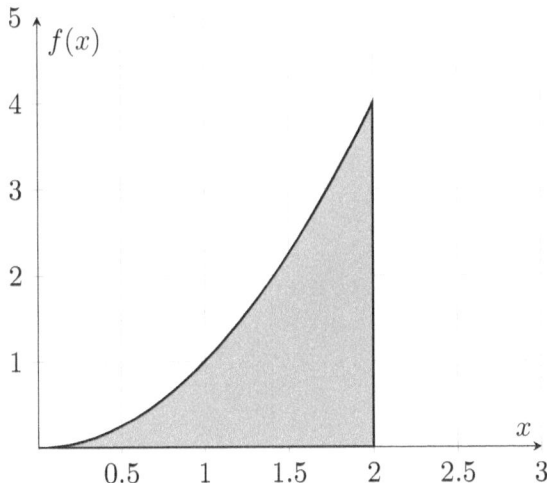

Figure 8.2: Shaded region showing the area under the curve $f(x) = x^2$ from $x = 0$ to $x = 2$.

Solving Problems of Accumulation

Integrals can also be used to solve problems of accumulation, such as finding the total distance traveled by an object if we know its velocity function.

Example: Accumulation Problem

An object moves with velocity $v(t) = 3t^2$ meters per second. Find the total distance traveled by the object from $t = 0$ to $t = 4$.
The total distance is the integral of the velocity function:

$$\text{Distance} = \int_0^4 3t^2 \, dt = \left[t^3\right]_0^4 = 4^3 - 0 = 64 \text{ meters}$$

Recap

In this chapter, we introduced the key concepts of calculus: derivatives and integrals. We explored the derivative as a measure of rates of change, covering basic differentiation rules and real-world applications such as motion and related rates. We also introduced the integral as a tool for calculating areas and solving accumulation problems. These foundational topics in calculus will provide a strong basis for further study in mathematics and its applications.

Chapter 9

Problem-Solving Strategies

Solving math problems requires both understanding and strategy. In this chapter, we'll explore general techniques for solving mathematical problems, real-world applications of math concepts, and offer practice problems with detailed solutions to solidify your understanding of all the topics covered in this book.

9.1 General Problem-Solving Techniques

Approaching math problems methodically is key to success. In this section, we will explore general techniques that apply to a variety of mathematical problems and discuss common pitfalls and how to avoid them.

9.1.1 How to Approach Math Problems Methodically

When faced with a math problem, a structured approach can lead to effective problem-solving. Here are some steps to consider:

- **Understand the Problem:** Read the problem carefully and identify what is being asked. Break it down into smaller parts if necessary.

- **Devise a Plan:** Think about the concepts and formulas that might help solve the problem. Consider similar problems you've encountered before.

- **Carry Out the Plan:** Execute your solution methodically. If the solution involves multiple steps, tackle each step one at a time.

- **Review Your Solution:** Check your answer to ensure it makes sense and that you've answered the question fully.

Example: Applying a Methodical Approach

Let's solve the quadratic equation $2x^2 - 4x - 6 = 0$ using the quadratic formula.

- Step 1: Identify the coefficients $a = 2$, $b = -4$, and $c = -6$.
- Step 2: Write the quadratic formula:

$$x = \frac{-b \pm \sqrt{b^2 - 4ac}}{2a}$$

- Step 3: Substitute the values into the formula:

$$x = \frac{-(-4) \pm \sqrt{(-4)^2 - 4(2)(-6)}}{2(2)} = \frac{4 \pm \sqrt{16 + 48}}{4} = \frac{4 \pm \sqrt{64}}{4} = \frac{4 \pm 8}{4}$$

- Step 4: Simplify the solutions:

$$x = \frac{4 + 8}{4} = 3 \quad \text{or} \quad x = \frac{4 - 8}{4} = -1$$

Thus, the solutions are $x = 3$ and $x = -1$.

9.1.2 Common Pitfalls and How to Avoid Them

Many mistakes in solving math problems come from misunderstandings or overlooking simple steps. Here are some common pitfalls and strategies to avoid them:

- **Misinterpreting the Problem:** Always take the time to fully understand the problem before attempting to solve it.

- **Forgetting to Check Units:** When solving word problems, always check your units and ensure they are consistent.

- **Overcomplicating the Solution:** Sometimes, a problem can be solved more easily than it first appears. Look for simpler methods before diving into complex ones.

- **Skipping Steps:** Avoid skipping steps in multi-step problems. Writing out each step clearly helps to avoid errors.

9.2 Real-World Math Problems

Mathematics isn't just confined to textbooks; it's used to solve real-world problems in various fields such as engineering, economics, and the natural sciences. In this section, we'll explore how mathematical concepts are applied to practical problems, using case studies and examples.

9.2.1 Applying Mathematical Concepts to Real-Life Situations

Mathematics can be used to solve real-life problems in finance, physics, engineering, and beyond. Here are some examples where math is crucial:

Example 1: Compound Interest

In finance, compound interest is calculated using the formula:

$$A = P\left(1 + \frac{r}{n}\right)^{nt}$$

Where:

- A is the amount of money accumulated after n years, including interest.

- P is the principal amount (the initial investment).

- r is the annual interest rate (as a decimal).
- n is the number of times that interest is compounded per year.
- t is the number of years the money is invested or borrowed for.

Case Study:

Suppose you invest $1,000 at an annual interest rate of 5% compounded quarterly. How much will you have after 10 years?
Using the formula:

$$A = 1000\left(1 + \frac{0.05}{4}\right)^{4(10)} = 1000\,(1.0125)^{40} = 1000 \times 1.643619 = 1643.62$$

After 10 years, you will have $1,643.62.

Example 2: Projectile Motion

In physics, the motion of an object under the influence of gravity can be described by the equation:

$$h(t) = -\frac{1}{2}gt^2 + v_0 t + h_0$$

Where:

- $h(t)$ is the height of the object at time t.
- g is the acceleration due to gravity (9.8 m/s²).
- v_0 is the initial velocity.
- h_0 is the initial height.

Case Study:

If a ball is thrown upward with an initial velocity of 15 m/s from a height of 2 m, when will the ball hit the ground?
Solving $h(t) = 0$ gives us the time when the ball hits the ground:

$$0 = -4.9t^2 + 15t + 2$$

Using the quadratic formula:

$$t = \frac{-15 \pm \sqrt{15^2 - 4(-4.9)(2)}}{2(-4.9)} = \frac{-15 \pm \sqrt{225 + 39.2}}{-9.8} = \frac{-15 \pm \sqrt{264.2}}{-9.8}$$

$$t = \frac{-15 \pm 16.25}{-9.8}$$

Taking the positive root:

$$t = \frac{-15 + 16.25}{-9.8} = 0.13 \text{ seconds}$$

9.3 Practice Problems and Solutions

Practice is essential to mastering mathematics. Below is a selection of practice problems covering topics from this book. Each problem is followed by a detailed solution to guide you through the process.

9.3.1 Problem 1: Solving Systems of Equations

Solve the following system of equations using the substitution method:

$$2x + 3y = 12$$
$$x - y = 4$$

Solution:

- Step 1: Solve the second equation for x:

$$x = y + 4$$

- Step 2: Substitute $x = y + 4$ into the first equation:

$$2(y + 4) + 3y = 12$$

$$2y + 8 + 3y = 12 \quad \Rightarrow \quad 5y + 8 = 12$$

$$5y = 4 \quad \Rightarrow \quad y = \frac{4}{5}$$

- Step 3: Substitute $y = \frac{4}{5}$ into $x = y + 4$:

$$x = \frac{4}{5} + 4 = \frac{4}{5} + \frac{20}{5} = \frac{24}{5}$$

Thus, the solution is $x = \frac{24}{5}$ and $y = \frac{4}{5}$.

9.3.2 Problem 2: Finding the Derivative

Find the derivative of the function $f(x) = 5x^3 - 2x^2 + x - 7$.

Solution:

Using the power rule for each term, we get:

$$f'(x) = 15x^2 - 4x + 1$$

9.3.3 Problem 3: Calculating Area Under a Curve

Find the area under the curve $f(x) = x^2$ from $x = 0$ to $x = 3$.

Solution:

The area under the curve is the definite integral of $f(x) = x^2$ from 0 to 3:

$$\int_0^3 x^2 \, dx = \left[\frac{x^3}{3}\right]_0^3 = \frac{27}{3} - 0 = 9$$

Thus, the area under the curve is 9 square units.

Recap

In this chapter, we explored general problem-solving techniques, applying mathematical concepts to real-world problems, and provided a series of practice problems with detailed solutions. By mastering these strategies and practicing consistently, you can improve your mathematical problem-solving skills and gain confidence in tackling complex problems.

Chapter 10

Tools and Resources for Continued Learning

Mathematics is a lifelong journey of learning and discovery. With the right tools and resources, you can continue developing your math skills beyond this book. In this chapter, we will explore how technology can enhance your math learning, provide a curated list of trustworthy resources for further study, and offer strategies for preparing for exams and applying your math skills in real-world settings.

10.1 Using Technology in Math

Technology has revolutionized the way we learn and apply mathematics. From simple calculators to advanced software, technology can help you deepen your understanding of mathematical concepts, solve complex problems efficiently, and visualize mathematical data.

10.1.1 Introduction to Calculators, Software, and Online Resources

Here are some essential tools you can use to enhance your math learning:

- **Calculators:**

- **Scientific Calculators:** Basic scientific calculators, such as the TI-30X or Casio fx-115, are great for solving algebraic equations, performing trigonometric functions, and more.
- **Graphing Calculators:** Graphing calculators, such as the TI-84 Plus or Casio fx-9750GII, allow you to plot functions, explore graphs, and solve more complex equations.

- **Mathematical Software:**
  - **WolframAlpha:** A powerful online tool that can solve equations, generate graphs, perform calculus operations, and more. WolframAlpha is useful for both students and professionals.
  - **GeoGebra:** Free and interactive software for graphing, geometry, and algebra. It's a great tool for visualizing mathematical concepts and solving problems interactively.
  - **MATLAB/Octave:** MATLAB is a high-level programming language and environment for solving mathematical and engineering problems. Octave is an open-source alternative that offers similar functionality.

- **Online Resources:**
  - **Khan Academy:** A free online resource offering comprehensive lessons in mathematics, from elementary math to advanced calculus and statistics. The platform includes instructional videos, practice problems, and quizzes.
  - **Desmos:** An online graphing calculator that is intuitive and free to use. Desmos helps you explore functions and graphs interactively.

10.1.2 How to Effectively Use Technology to Enhance Learning

Technology is most effective when used as a complement to your learning process, not a substitute for understanding the concepts. Here are some strategies for making the best use of technology in your math studies:

- **Use Technology to Check Your Work:** After solving a problem manually, use software like WolframAlpha or a graphing calculator to verify your results. This will help reinforce your understanding of the concepts.

- **Visualize Complex Concepts:** Use tools like GeoGebra or Desmos to visualize functions, graphs, and geometric shapes. Seeing math visually can deepen your comprehension.

- **Practice Interactively:** Take advantage of online resources like Khan Academy to engage in interactive lessons and quizzes, which can provide instant feedback on your progress.

10.2 Additional Learning Resources

In this section, I provide you with reliable and trustworthy resources to further your study of mathematics. Whether you're looking for books, websites, or courses, these resources will help you expand your knowledge and master more advanced concepts.

10.2.1 Books for Further Study

Here are some highly recommended books for students at various levels of mathematical study:

- **Precalculus: Mathematics for Calculus** by James Stewart, Lothar Redlin, and Saleem Watson – A comprehensive text covering topics essential for calculus, including functions, trigonometry, and analytical geometry.

- **Calculus: Early Transcendentals** by James Stewart – One of the most popular and well-respected calculus textbooks. It covers differentiation, integration, and applications of calculus in a clear and accessible manner.

- **Introduction to Linear Algebra** by Gilbert Strang – This textbook is highly regarded for its clear explanations and practical applications in linear algebra, making it an excellent resource for students advancing to higher-level mathematics.

- **How to Solve It** by George Pólya – A classic text on problem-solving strategies in mathematics. This book provides a step-by-step framework for approaching and solving a wide variety of math problems.

10.2.2 Websites and Online Resources

In addition to the tools mentioned earlier, here are more websites and platforms to continue your math education:

- **Brilliant.org:** An interactive platform offering courses in math, science, and computer science. Brilliant is known for its engaging problem-solving approach to learning.

- **Paul's Online Math Notes:** A comprehensive collection of math notes covering algebra, calculus, differential equations, and linear algebra. This site is a great resource for reviewing key concepts and examples.

- **MIT OpenCourseWare (OCW):** A free online resource offering materials from a wide range of MIT courses. You can access lecture notes, assignments, and exams from math courses at one of the top universities in the world.

- **Coursera:** An online learning platform offering courses in mathematics from top universities. Coursera offers both free and paid options, and you can earn certificates from institutions such as Stanford, University of Michigan, and more.

10.2.3 Courses and Tutorials

For those who want to dive deeper into specific areas of mathematics, the following platforms offer structured courses and tutorials:

- **edX:** edX offers online courses from universities like MIT, Harvard, and Berkeley. You can find courses in algebra, calculus, data science, and more.

- **FutureLearn:** FutureLearn provides free and paid courses from top universities and organizations. It's an excellent resource for building specialized knowledge in mathematics.

- **YouTube Channels:** Channels like 3Blue1Brown, Numberphile, and Mathologer offer engaging and insightful mathematical videos that explore both fundamental and advanced concepts.

10.2.4 Tips for Finding and Utilizing Resources Effectively

Here are a few tips to help you make the most of the resources available to you:

- **Prioritize Quality over Quantity:** Focus on a few reliable resources and engage deeply with them, rather than spreading yourself too thin across too many materials.

- **Use Practice Problems to Reinforce Learning:** Whether you're using books, websites, or online courses, always seek out practice problems to test your understanding and solidify concepts.

- **Track Your Progress:** Keep a log of topics you've studied and practice problems you've solved. This will help you identify areas for improvement and ensure you're making steady progress.

10.3 Preparing for Exams and Beyond

Preparing for math exams—whether standardized tests or college entrance exams—requires a strategic approach. In this section, I provide study strategies and insights on how to apply your math skills in academic and professional settings.

10.3.1 Study Strategies for Standardized Tests and College Entrance Exams

Success on standardized tests, such as the SAT, ACT, or AP Calculus exam, requires both content knowledge and test-taking strategies. Here are some tips to help you prepare:

- **Understand the Test Format:** Familiarize yourself with the format of the test, including the types of questions asked and the time limits. This will help you manage your time effectively during the exam.

- **Focus on Weak Areas:** Identify which topics you struggle with the most and dedicate extra time to practicing those areas. Use resources such as practice tests to identify your weaknesses.

- **Practice Under Timed Conditions:** Simulate exam conditions by practicing under timed constraints. This will help you improve your speed and accuracy when it comes to solving problems under pressure.

- **Use Official Practice Tests:** Many testing organizations provide free official practice tests. Make use of these resources to ensure you're practicing with the most accurate materials.

10.3.2 How to Apply Math Skills in Academic and Professional Settings

Math skills are not just for passing exams—they are invaluable in many academic and professional settings. Here are some ways you can apply your math skills beyond the classroom:

- **Academic Research:** In fields such as engineering, economics, and the natural sciences, math is essential for conducting research, analyzing data, and developing models.

- **Business and Finance:** Whether you're managing budgets, analyzing financial reports, or making investment decisions, math is a crucial tool in business and finance.

- **Problem-Solving in Everyday Life:** Math skills can help you make informed decisions in your personal life, from managing your finances to interpreting statistics in the news.

Recap

In this chapter, we explored the importance of technology in enhancing your math learning, providing a range of tools from calculators to software. We also offered a curated list of additional learning resources to continue your education and shared strategies for preparing for exams and applying math in academic and professional settings. With these tools and strategies, you are well-equipped to continue your journey in mathematics.

Conclusion

Recap of Key Concepts

Over the course of this book, we've explored a wide range of mathematical topics, each building on the previous to form a complete foundation for understanding advanced math. Let's briefly recap some of the most important concepts covered in each chapter:

- **Chapter 1: Algebra Fundamentals** – We started with the basics of algebra, including solving equations, understanding inequalities, and manipulating algebraic expressions.

- **Chapter 2: Quadratic Equations and Functions** – You learned how to solve quadratic equations by factoring, completing the square, and using the quadratic formula, as well as how to graph quadratic functions.

- **Chapter 3: Graphing and Functions** – We explored the relationship between equations and graphs, learning to visualize and interpret various types of functions.

- **Chapter 4: Geometry and 3D Shapes** – This chapter introduced geometric principles, the properties of shapes, and volume calculations for 3D objects.

- **Chapter 5: Trigonometry Without Tears** – We tackled trigonometric ratios, the unit circle, and basic trigonometric identities, laying the groundwork for advanced applications.

- **Chapter 6: Probability and Statistics** – You explored probability rules and learned how to analyze and interpret data using descriptive statistics.

- **Chapter 7: Advanced Algebra and Pre-Calculus** – This chapter focused on sequences, series, matrices, and limits, bridging the gap between algebra and calculus.

- **Chapter 8: Introduction to Calculus** – We introduced the fundamental concepts of differentiation and integration, highlighting their real-world applications.

- **Chapter 9: Problem-Solving Strategies** – You learned how to approach math problems systematically, avoid common pitfalls, and apply mathematical concepts to real-world problems.

- **Chapter 10: Tools and Resources for Continued Learning** – Finally, we explored the role of technology in learning mathematics and provided you with reliable resources to continue your journey in math.

These chapters have equipped you with a strong foundation in mathematics. But remember, this is just the beginning. Math is a vast and ever-evolving field, and there is always more to learn.

Encouragement for Further Study

Mathematics is not just about solving problems—it's about developing critical thinking skills, creativity, and persistence. Throughout your mathematical journey, you will encounter challenges that require you to push your boundaries and think in new ways. Don't be discouraged by difficulties; instead, see them as opportunities to grow and deepen your understanding.

We encourage you to continue exploring mathematics beyond this book. Whether you are preparing for college, entering a math-related career, or simply looking to enrich your understanding of the world, the skills you've learned here will serve as a solid foundation. Use the tools and resources provided in this book to further your studies and continue mastering the beautiful language of math.

Remember that mathematics is a lifelong journey. The more you explore, the more you'll discover how math connects to various aspects of life—from the sciences and engineering to economics and even the arts.

Final Thoughts

As we conclude this book, let's take a moment to reflect on the role of mathematics in both personal and professional growth.

Math teaches us more than just how to calculate numbers—it fosters a way of thinking that can be applied to all areas of life. Whether it's problem-solving, logical reasoning, or analytical thinking, the skills developed through studying math are invaluable in both personal and professional contexts.

In your personal life, math helps you make informed decisions, manage finances, and approach everyday challenges with a logical mindset. In your professional life, whether you pursue a career in engineering, finance, computer science, or any field that requires data-driven decision-making, math will continue to play a pivotal role in your success.

I hope this book has inspired you to not only appreciate the power of math but also to recognize its importance in the world around you. Remember that every problem you solve, every concept you master, and every new challenge you take on is a step toward a greater understanding of how math shapes our world.

Thank you for joining me on this journey through mathematics. I wish you continued success as you unlock new doors and discover even more exciting possibilities in the world of math.

Final Encouragement

Mathematics is an adventure that never ends. With persistence, curiosity, and the right resources, you can go as far as your interests take

CONCLUSION

you. Keep exploring, keep solving, and, most importantly, keep enjoying the journey.

About the Author

Fernando Abednego Halim, a distinguished author from Surabaya, Indonesia, beautifully combines his deep knowledge of physics with his talent for storytelling. Educated at Universität Hamburg and currently based in Jakarta, he gently demystifies complex physics concepts for his readers, making them both engaging and accessible. His writings thoughtfully lead readers through the mysterious aspects of quantum phenomena, turning intricate subjects into something clear and inviting.

In addition to his expertise in physics, Fernando has a profound passion for mathematics. Though not a mathematician by profession, he has been a dedicated math tutor for many years. His love for numbers and problem-solving has inspired him to help students of all levels grasp mathematical concepts with clarity and confidence. Whether simplifying algebra or explaining the intricacies of calculus, Fernando's approach to teaching math reflects his belief that everyone can succeed with the right guidance.

Recently, Fernando has begun exploring the realms of artificial intelligence and generative AI, enriching his academic pursuits. This new exploration invites his readers to journey with him into the unfolding potentials and nuances of AI, continuing to nurture their curiosity for both established and newly emerging scientific fields.

Fernando's work exemplifies his commitment to making complex scientific and mathematical ideas accessible to everyone, fostering a deeper appreciation for these subjects in both academic and everyday contexts.

ABOUT THE AUTHOR

You can connect with Fernando on LinkedIn at:
https://www.linkedin.com/in/fernando-abednego-halim

www.ingramcontent.com/pod-product-compliance
Lightning Source LLC
Chambersburg PA
CBHW030446220526

45464CB00006B/2428